DETAILS

Name:	
Company:	
Contact Number:	
Contact Number:	
Email:	

Emergency Details:

Name:		Name:	
Company:		Company:	

Important:

Date:		Day:	Mon	Tue	Wed	Thu	Fri	Sat	Sun
Foreman:									
Contract No:									

Hours Lost Due To Bad Weather	Visitors

Weather Conditions		
AM	PM	

Schedule	Problems / Delays
Completion Date:	
Days Ahead of Schedute:	
Days Behind Schedute:	

Safely Issues:	Accidents/Incidents:

Summary Of Work Performed Today

Signature:	Name:

Equipment On Site	No Units	Working	
		Yes	No

Employee /Contractor	Trade	Contracted Hours	Overtime

Materials Delivered	From & Rate	Equipment Rented	No Units

Notes:

Date:		**Day:**	Mon	Tue	Wed	Thu	Fri	Sat	Sun
Foreman:									
Contract No:									

Hours Lost Due To Bad Weather	Visitors

Weather Conditions	
AM	PM

Schedule	Problems / Delays
Completion Date:	
Days Ahead of Schedute:	
Days Behind Schedute:	

Safely Issues:	Accidents/Incidents:

Summary Of Work Performed Today

Signature:	Name:

Equipment On Site	No Units	Working	
		Yes	No

Employee /Contractor	Trade	Contracted Hours	Overtime

Materials Delivered	From & Rate	Equipment Rented	No Units

Notes:

Date:		Day:	Mon	Tue	Wed	Thu	Fri	Sat	Sun
Foreman:									
Contract No:									

Hours Lost Due To Bad Weather	Visitors

Weather Conditions

AM	PM	

Schedule	Problems / Delays
Completion Date:	
Days Ahead of Schedule:	
Days Behind Schedute:	

Safely Issues:	Accidents/Incidents:

Summary Of Work Performed Today

Signature:	Name:

Equipment On Site	No Units	Working	
		Yes	No

Employee /Contractor	Trade	Contracted Hours	Overtime

Materials Delivered	From & Rate	Equipment Rented	No Units

Notes:

Date:		Day:	Mon	Tue	Wed	Thu	Fri	Sat	Sun
Foreman:									
Contract No:									

Hours Lost Due To Bad Weather	Visitors

Weather Conditions		Visitors (cont.)
AM	PM	

Schedule	Problems / Delays
Completion Date:	
Days Ahead of Schedute:	
Days Behind Schedule:	

Safely Issues:	Accidents/Incidents:

Summary Of Work Performed Today

Signature:	Name:

Equipment On Site	No Units	Working	
		Yes	No

Employee /Contractor	Trade	Contracted Hours	Overtime

Materials Delivered	From & Rate	Equipment Rented	No Units

Notes:

Date:		**Day:**	Mon	Tue	Wed	Thu	Fri	Sat	Sun

Foreman:	
Contract No:	

Hours Lost Due To Bad Weather	Visitors

Weather Conditions

AM	PM

Schedule	Problems / Delays
Completion Date:	
Days Ahead of Schedute:	
Days Behind Schedule:	

Safely Issues:	Accidents/Incidents:

Summary Of Work Performed Today

Signature:	Name:

Equipment On Site	No Units	Working	
		Yes	No

Employee /Contractor	Trade	Contracted Hours	Overtime

Materials Delivered	From & Rate	Equipment Rented	No Units

Notes:

Date:		Day:	Mon	Tue	Wed	Thu	Fri	Sat	Sun
Foreman:									
Contract No:									

Hours Lost Due To Bad Weather	Visitors

Weather Conditions	
AM	PM

Schedule	Problems / Delays
Completion Date:	
Days Ahead of Schedute:	
Days Behind Schedule:	

Safely Issues:	Accidents/Incidents:

Summary Of Work Performed Today

Signature:	Name:

Equipment On Site	No Units	Working	
		Yes	No

Employee /Contractor	Trade	Contracted Hours	Overtime

Materials Delivered	From & Rate	Equipment Rented	No Units

Notes:

Date:		Day:	Mon	Tue	Wed	Thu	Fri	Sat	Sun
Foreman:									
Contract No:									

Hours Lost Due To Bad Weather	Visitors

Weather Conditions	
AM	PM

Schedule	Problems / Delays
Completion Date:	
Days Ahead of Schedute:	
Days Behind Schedule:	

Safely Issues:	Accidents/Incidents:

Summary Of Work Performed Today

Signature:	Name:

Equipment On Site	No Units	Working	
		Yes	No

Employee /Contractor	Trade	Contracted Hours	Overtime

Materials Delivered	From & Rate	Equipment Rented	No Units

Notes:

Date:		**Day:**	Mon	Tue	Wed	Thu	Fri	Sat	Sun
Foreman:									
Contract No:									

Hours Lost Due To Bad Weather	Visitors

Weather Conditions		
AM	PM	

Schedule	Problems / Delays
Completion Date:	
Days Ahead of Schedule:	
Days Behind Schedute:	

Safely Issues:	Accidents/Incidents:

Summary Of Work Performed Today

Signature:	Name:

Equipment On Site	No Units	Working	
		Yes	No

Employee /Contractor	Trade	Contracted Hours	Overtime

Materials Delivered	From & Rate	Equipment Rented	No Units

Notes:

Date:		**Day:**	Mon	Tue	Wed	Thu	Fri	Sat	Sun
Foreman:									
Contract No:									

Hours Lost Due To Bad Weather	Visitors

Weather Conditions

AM	PM	

Schedule	Problems / Delays
Completion Date:	
Days Ahead of Schedute:	
Days Behind Schedute:	

Safely Issues:	Accidents/Incidents:

Summary Of Work Performed Today

Signature:	Name:

Equipment On Site	No Units	Working	
		Yes	No

Employee /Contractor	Trade	Contracted Hours	Overtime

Materials Delivered	From & Rate	Equipment Rented	No Units

Notes:

Date:		**Day:**	Mon	Tue	Wed	Thu	Fri	Sat	Sun
Foreman:									
Contract No:									

Hours Lost Due To Bad Weather	Visitors

Weather Conditions	
AM	PM

Schedule

Completion Date:	
Days Ahead of Schedule:	
Days Behind Schedute:	

Problems / Delays

Safely Issues:

Accidents/Incidents:

Summary Of Work Performed Today

Signature:	Name:

Equipment On Site	No Units	Working	
		Yes	No

Employee /Contractor	Trade	Contracted Hours	Overtime

Materials Delivered	From & Rate	Equipment Rented	No Units

Notes:

Date:		Day:	Mon	Tue	Wed	Thu	Fri	Sat	Sun
Foreman:									
Contract No:									

Hours Lost Due To Bad Weather	Visitors

Weather Conditions

AM	PM	

Schedule	Problems / Delays
Completion Date:	
Days Ahead of Schedule:	
Days Behind Schedute:	

Safely Issues:	Accidents/Incidents:

Summary Of Work Performed Today

Signature:	Name:

Equipment On Site	No Units	Working	
		Yes	No

Employee /Contractor	Trade	Contracted Hours	Overtime

Materials Delivered	From & Rate	Equipment Rented	No Units

Notes:

Date:		Day:	Mon	Tue	Wed	Thu	Fri	Sat	Sun
Foreman:									
Contract No:									

Hours Lost Due To Bad Weather	Visitors

Weather Conditions	
AM	PM

Schedule	Problems / Delays
Completion Date:	
Days Ahead of Schedute:	
Days Behind Schedute:	

Safely Issues:	Accidents/Incidents:

Summary Of Work Performed Today

Signature:	Name:

Equipment On Site	No Units	Working	
		Yes	No

Employee /Contractor	Trade	Contracted Hours	Overtime

Materials Delivered	From & Rate	Equipment Rented	No Units

Notes:

Date:		**Day:**	Mon	Tue	Wed	Thu	Fri	Sat	Sun
Foreman:									
Contract No:									

Hours Lost Due To Bad Weather | Visitors

Hours Lost Due To Bad Weather	Visitors

Weather Conditions	
AM	PM

Schedule	Problems / Delays
Completion Date:	
Days Ahead of Schedule:	
Days Behind Schedule:	

Safely Issues:	Accidents/Incidents:

Summary Of Work Performed Today

Signature:	Name:

Equipment On Site	No Units	Working	
		Yes	No

Employee /Contractor	Trade	Contracted Hours	Overtime

Materials Delivered	From & Rate	Equipment Rented	No Units

Notes:

Date:		**Day:**	Mon	Tue	Wed	Thu	Fri	Sat	Sun
Foreman:									
Contract No:									

Hours Lost Due To Bad Weather	Visitors

Weather Conditions	
AM	PM

Schedule	Problems / Delays
Completion Date:	
Days Ahead of Schedute:	
Days Behind Schedute:	

Safely Issues:	Accidents/Incidents:

Summary Of Work Performed Today

Signature:	Name:

Equipment On Site	No Units	Working	
		Yes	No

Employee /Contractor	Trade	Contracted Hours	Overtime

Materials Delivered	From & Rate	Equipment Rented	No Units

Notes:

Date:		**Day:**	Mon	Tue	Wed	Thu	Fri	Sat	Sun
Foreman:									
Contract No:									

Hours Lost Due To Bad Weather	Visitors

Weather Conditions	
AM	PM

Schedule	Problems / Delays
Completion Date:	
Days Ahead of Schedule:	
Days Behind Schedute:	

Safely Issues:	Accidents/Incidents:

Summary Of Work Performed Today

Signature:	Name:

Equipment On Site	No Units	Working	
		Yes	No

Employee /Contractor	Trade	Contracted Hours	Overtime

Materials Delivered	From & Rate	Equipment Rented	No Units

Notes:

Date:		**Day:**	Mon	Tue	Wed	Thu	Fri	Sat	Sun
Foreman:									
Contract No:									

Hours Lost Due To Bad Weather	Visitors

Weather Conditions	
AM	PM

Schedule	Problems / Delays
Completion Date:	
Days Ahead of Schedute:	
Days Behind Schedule:	

Safely Issues:	Accidents/Incidents:

Summary Of Work Performed Today

Signature:	Name:

Equipment On Site	No Units	Working	
		Yes	No

Employee /Contractor	Trade	Contracted Hours	Overtime

Materials Delivered	From & Rate	Equipment Rented	No Units

Notes:

Date:		Day:	Mon Tue Wed Thu Fri Sat Sun
Foreman:			
Contract No:			

Hours Lost Due To Bad Weather	Visitors

Weather Conditions		
AM	PM	

Schedule	Problems / Delays
Completion Date:	
Days Ahead of Schedute:	
Days Behind Schedute:	

Safely Issues:	Accidents/Incidents:

Summary Of Work Performed Today

Signature:	Name:

Equipment On Site	No Units	Working	
		Yes	No

Employee /Contractor	Trade	Contracted Hours	Overtime

Materials Delivered	From & Rate	Equipment Rented	No Units

Notes:

Date:		**Day:**	Mon	Tue	Wed	Thu	Fri	Sat	Sun
Foreman:									
Contract No:									

Hours Lost Due To Bad Weather	Visitors

Weather Conditions		
AM	PM	

Schedule	Problems / Delays
Completion Date:	
Days Ahead of Schedule:	
Days Behind Schedute:	

Safely Issues:	Accidents/Incidents:

Summary Of Work Performed Today

Signature:	Name:

Equipment On Site	No Units	Working	
		Yes	No

Employee /Contractor	Trade	Contracted Hours	Overtime

Materials Delivered	From & Rate	Equipment Rented	No Units

Notes:

Date:		Day:	Mon	Tue	Wed	Thu	Fri	Sat	Sun
Foreman:									
Contract No:									

Hours Lost Due To Bad Weather	Visitors

Weather Conditions		
AM	PM	

Schedule		Problems / Delays
Completion Date:		
Days Ahead of Schedule:		
Days Behind Schedute:		

Safely Issues:	Accidents/Incidents:

Summary Of Work Performed Today

Signature:	Name:

Equipment On Site	No Units	Working	
		Yes	No

Employee /Contractor	Trade	Contracted Hours	Overtime

Materials Delivered	From & Rate	Equipment Rented	No Units

Notes:

Date:		Day:	Mon	Tue	Wed	Thu	Fri	Sat	Sun
Foreman:									
Contract No:									

Hours Lost Due To Bad Weather	Visitors

Weather Conditions	
AM	PM

Schedule	Problems / Delays
Completion Date:	
Days Ahead of Schedule:	
Days Behind Schedute:	

Safely Issues:	Accidents/Incidents:

Summary Of Work Performed Today

Signature:	Name:

Equipment On Site	No Units	Working	
		Yes	No

Employee /Contractor	Trade	Contracted Hours	Overtime

Materials Delivered	From & Rate	Equipment Rented	No Units

Notes:

Date:		**Day:**	Mon	Tue	Wed	Thu	Fri	Sat	Sun
Foreman:									
Contract No:									

Hours Lost Due To Bad Weather	Visitors

Weather Conditions	
AM	PM

Schedule	Problems / Delays
Completion Date:	
Days Ahead of Schedute:	
Days Behind Schedute:	

Safely Issues:	Accidents/Incidents:

Summary Of Work Performed Today

Signature:	Name:

Equipment On Site	No Units	Working	
		Yes	No

Employee /Contractor	Trade	Contracted Hours	Overtime

Materials Delivered	From & Rate	Equipment Rented	No Units

Notes:

Date:		**Day:** Mon Tue Wed Thu Fri Sat Sun
Foreman:		
Contract No:		

Hours Lost Due To Bad Weather	Visitors

Weather Conditions	
AM	PM

Schedule	Problems / Delays
Completion Date:	
Days Ahead of Schedute:	
Days Behind Schedute:	

Safely Issues:	Accidents/Incidents:

Summary Of Work Performed Today

Signature:	Name:

Equipment On Site	No Units	Working	
		Yes	No

Employee /Contractor	Trade	Contracted Hours	Overtime

Materials Delivered	From & Rate	Equipment Rented	No Units

Notes:

Date:		Day:	Mon	Tue	Wed	Thu	Fri	Sat	Sun
Foreman:									
Contract No:									

Hours Lost Due To Bad Weather	Visitors

Weather Conditions	
AM	PM

Schedule	Problems / Delays
Completion Date:	
Days Ahead of Schedule:	
Days Behind Schedute:	

Safely Issues:	Accidents/Incidents:

Summary Of Work Performed Today

Signature:	Name:

Equipment On Site	No Units	Working	
		Yes	No

Employee /Contractor	Trade	Contracted Hours	Overtime

Materials Delivered	From & Rate	Equipment Rented	No Units

Notes:

Date:		**Day:**	Mon	Tue	Wed	Thu	Fri	Sat	Sun
Foreman:									
Contract No:									

Hours Lost Due To Bad Weather	Visitors

Weather Conditions		
AM	PM	

Schedule	Problems / Delays
Completion Date:	
Days Ahead of Schedule:	
Days Behind Schedule:	

Safely Issues:	Accidents/Incidents:

Summary Of Work Performed Today

Signature:	Name:

Equipment On Site	No Units	Working	
		Yes	No

Employee /Contractor	Trade	Contracted Hours	Overtime

Materials Delivered	From & Rate	Equipment Rented	No Units

Notes:

Date:		**Day:**	Mon	Tue	Wed	Thu	Fri	Sat	Sun
Foreman:									
Contract No:									

Hours Lost Due To Bad Weather

Weather Conditions

AM	PM

Visitors

Schedule

Completion Date:	
Days Ahead of Schedute:	
Days Behind Schedule:	

Problems / Delays

Safely Issues:

Accidents/Incidents:

Summary Of Work Performed Today

Signature:	Name:

Equipment On Site	No Units	Working	
		Yes	No

Employee /Contractor	Trade	Contracted Hours	Overtime

Materials Delivered	From & Rate	Equipment Rented	No Units

Notes:

Date:		Day:	Mon Tue Wed Thu Fri Sat Sun
Foreman:			
Contract No:			

Hours Lost Due To Bad Weather

Visitors

Weather Conditions

AM	PM

Schedule

Completion Date:	
Days Ahead of Schedute:	
Days Behind Schedute:	

Problems / Delays

Safely Issues:

Accidents/Incidents:

Summary Of Work Performed Today

Signature:	Name:

Equipment On Site	No Units	Working	
		Yes	No

Employee /Contractor	Trade	Contracted Hours	Overtime

Materials Delivered	From & Rate	Equipment Rented	No Units

Notes:

Date:		Day:	Mon	Tue	Wed	Thu	Fri	Sat	Sun
Foreman:									
Contract No:									

Hours Lost Due To Bad Weather	Visitors

Weather Conditions		
AM	PM	

Schedule	Problems / Delays
Completion Date:	
Days Ahead of Schedute:	
Days Behind Schedute:	

Safely Issues:	Accidents/Incidents:

Summary Of Work Performed Today

Signature:	Name:

Equipment On Site	No Units	Working	
		Yes	No

Employee /Contractor	Trade	Contracted Hours	Overtime

Materials Delivered	From & Rate	Equipment Rented	No Units

Notes:

Date:		**Day:** Mon Tue Wed Thu Fri Sat Sun
Foreman:		
Contract No:		

Hours Lost Due To Bad Weather	Visitors

Weather Conditions	
AM	PM

Schedule	Problems / Delays
Completion Date:	
Days Ahead of Schedute:	
Days Behind Schedule:	

Safely Issues:	Accidents/Incidents:

Summary Of Work Performed Today

Signature:	Name:

Equipment On Site	No Units	Working	
		Yes	No

Employee /Contractor	Trade	Contracted Hours	Overtime

Materials Delivered	From & Rate	Equipment Rented	No Units

Notes:

Date:		**Day:**	Mon	Tue	Wed	Thu	Fri	Sat	Sun
Foreman:									
Contract No:									

Hours Lost Due To Bad Weather	Visitors

Weather Conditions		
AM	PM	

Schedule	Problems / Delays
Completion Date:	
Days Ahead of Schedute:	
Days Behind Schedule:	

Safely Issues:	Accidents/Incidents:

Summary Of Work Performed Today

Signature:	Name:

Equipment On Site	No Units	Working	
		Yes	No

Employee /Contractor	Trade	Contracted Hours	Overtime

Materials Delivered	From & Rate	Equipment Rented	No Units

Notes:

Date:		Day:	Mon	Tue	Wed	Thu	Fri	Sat	Sun
Foreman:									
Contract No:									

Hours Lost Due To Bad Weather	Visitors

Weather Conditions	
AM	PM

Schedule	Problems / Delays
Completion Date:	
Days Ahead of Schedule:	
Days Behind Schedute:	

Safely Issues:	Accidents/Incidents:

Summary Of Work Performed Today

Signature:	Name:

Equipment On Site	No Units	Working	
		Yes	No

Employee /Contractor	Trade	Contracted Hours	Overtime

Materials Delivered	From & Rate	Equipment Rented	No Units

Notes:

Date:		Day:	Mon	Tue	Wed	Thu	Fri	Sat	Sun
Foreman:									
Contract No:									

Hours Lost Due To Bad Weather	Visitors

Weather Conditions	
AM	PM

Schedule	Problems / Delays
Completion Date:	
Days Ahead of Schedule:	
Days Behind Schedule:	

Safely Issues:	Accidents/Incidents:

Summary Of Work Performed Today

Signature:	Name:

Equipment On Site	No Units	Working	
		Yes	No

Employee /Contractor	Trade	Contracted Hours	Overtime

Materials Delivered	From & Rate	Equipment Rented	No Units

Notes:

Date:		**Day:**	Mon	Tue	Wed	Thu	Fri	Sat	Sun
Foreman:									
Contract No:									

Hours Lost Due To Bad Weather	Visitors

Weather Conditions	
AM	PM

Schedule	Problems / Delays
Completion Date:	
Days Ahead of Schedute:	
Days Behind Schedute:	

Safely Issues:	Accidents/Incidents:

Summary Of Work Performed Today

Signature:	Name:

Equipment On Site	No Units	Working	
		Yes	No

Employee /Contractor	Trade	Contracted Hours	Overtime

Materials Delivered	From & Rate	Equipment Rented	No Units

Notes:

Date:		Day:	Mon	Tue	Wed	Thu	Fri	Sat	Sun
Foreman:									
Contract No:									

Hours Lost Due To Bad Weather	Visitors

Weather Conditions	
AM	PM

Schedule	Problems / Delays
Completion Date:	
Days Ahead of Schedute:	
Days Behind Schedute:	

Safely Issues:	Accidents/Incidents:

Summary Of Work Performed Today

Signature:	Name:

Equipment On Site	No Units	Working	
		Yes	No

Employee /Contractor	Trade	Contracted Hours	Overtime

Materials Delivered	From & Rate	Equipment Rented	No Units

Notes:

Date:		**Day:**	Mon	Tue	Wed	Thu	Fri	Sat	Sun
Foreman:									
Contract No:									

Hours Lost Due To Bad Weather	Visitors

Weather Conditions		
AM	PM	

Schedule		Problems / Delays
Completion Date:		
Days Ahead of Schedute:		
Days Behind Schedute:		

Safely Issues:	Accidents/Incidents:

Summary Of Work Performed Today

Signature:	Name:

Equipment On Site	No Units	Working	
		Yes	No

Employee /Contractor	Trade	Contracted Hours	Overtime

Materials Delivered	From & Rate	Equipment Rented	No Units

Notes:

Date:		**Day:**	Mon	Tue	Wed	Thu	Fri	Sat	Sun
Foreman:									
Contract No:									

Hours Lost Due To Bad Weather	Visitors

Weather Conditions		
AM	PM	

Schedule	Problems / Delays
Completion Date:	
Days Ahead of Schedule:	
Days Behind Schedute:	

Safely Issues:	Accidents/Incidents:

Summary Of Work Performed Today

Signature:	Name:

Equipment On Site	No Units	Working	
		Yes	No

Employee /Contractor	Trade	Contracted Hours	Overtime

Materials Delivered	From & Rate	Equipment Rented	No Units

Notes:

Date:		Day:	Mon	Tue	Wed	Thu	Fri	Sat	Sun
Foreman:									
Contract No:									

Hours Lost Due To Bad Weather	Visitors

Weather Conditions	
AM	PM

Schedule	Problems / Delays
Completion Date:	
Days Ahead of Schedule:	
Days Behind Schedule:	

Safely Issues:	Accidents/Incidents:

Summary Of Work Performed Today

Signature:	Name:

Equipment On Site	No Units	Working	
		Yes	No

Employee /Contractor	Trade	Contracted Hours	Overtime

Materials Delivered	From & Rate	Equipment Rented	No Units

Notes:

Date:		**Day:**	Mon	Tue	Wed	Thu	Fri	Sat	Sun
Foreman:									
Contract No:									

Hours Lost Due To Bad Weather	Visitors

Weather Conditions		
AM	PM	

Schedule	Problems / Delays
Completion Date:	
Days Ahead of Schedule:	
Days Behind Schedule:	

Safely Issues:	Accidents/Incidents:

Summary Of Work Performed Today

Signature:	Name:

Equipment On Site	No Units	Working	
		Yes	No

Employee /Contractor	Trade	Contracted Hours	Overtime

Materials Delivered	From & Rate	Equipment Rented	No Units

Notes:

Date:		**Day:**	Mon	Tue	Wed	Thu	Fri	Sat	Sun
Foreman:									
Contract No:									

Hours Lost Due To Bad Weather	Visitors

Weather Conditions			
AM	PM		

Schedule

	Problems / Delays
Completion Date:	
Days Ahead of Schedute:	
Days Behind Schedute:	

Safely Issues:	Accidents/Incidents:

Summary Of Work Performed Today

Signature:	Name:

Equipment On Site	No Units	Working	
		Yes	No

Employee /Contractor	Trade	Contracted Hours	Overtime

Materials Delivered	From & Rate	Equipment Rented	No Units

Notes:

Date:		**Day:**	Mon	Tue	Wed	Thu	Fri	Sat	Sun
Foreman:									
Contract No:									

Hours Lost Due To Bad Weather	Visitors

Weather Conditions

AM	PM

Schedule	Problems / Delays
Completion Date:	
Days Ahead of Schedute:	
Days Behind Schedute:	

Safely Issues:	Accidents/Incidents:

Summary Of Work Performed Today

Signature:	Name:

Equipment On Site	No Units	Working	
		Yes	No

Employee /Contractor	Trade	Contracted Hours	Overtime

Materials Delivered	From & Rate	Equipment Rented	No Units

Notes:

Date:		**Day:**	Mon	Tue	Wed	Thu	Fri	Sat	Sun
Foreman:									
Contract No:									

Hours Lost Due To Bad Weather	Visitors

Weather Conditions	
AM	PM

Schedule	Problems / Delays
Completion Date:	
Days Ahead of Schedute:	
Days Behind Schedule:	

Safely Issues:	Accidents/Incidents:

Summary Of Work Performed Today

Signature:	Name:

Equipment On Site	No Units	Working	
		Yes	No

Employee /Contractor	Trade	Contracted Hours	Overtime

Materials Delivered	From & Rate	Equipment Rented	No Units

Notes:

Date:		**Day:**	Mon	Tue	Wed	Thu	Fri	Sat	Sun
Foreman:									
Contract No:									

Hours Lost Due To Bad Weather

Visitors

Weather Conditions

AM	PM

Schedule

Completion Date:	
Days Ahead of Schedule:	
Days Behind Schedute:	

Problems / Delays

Safely Issues:

Accidents/Incidents:

Summary Of Work Performed Today

Signature:	Name:

Equipment On Site	No Units	Working	
		Yes	No

Employee /Contractor	Trade	Contracted Hours	Overtime

Materials Delivered	From & Rate	Equipment Rented	No Units

Notes:

Date:		**Day:**	Mon	Tue	Wed	Thu	Fri	Sat	Sun
Foreman:									
Contract No:									

Hours Lost Due To Bad Weather

Visitors

Weather Conditions

AM	PM

Schedule	Problems / Delays
Completion Date:	
Days Ahead of Schedute:	
Days Behind Schedule:	

Safely Issues:	Accidents/Incidents:

Summary Of Work Performed Today

Signature:	Name:

Equipment On Site	No Units	Working	
		Yes	No

Employee /Contractor	Trade	Contracted Hours	Overtime

Materials Delivered	From & Rate	Equipment Rented	No Units

Notes:

Date:		**Day:**	Mon	Tue	Wed	Thu	Fri	Sat	Sun
Foreman:									
Contract No:									

Hours Lost Due To Bad Weather	Visitors

Weather Conditions	
AM	PM

Schedule	Problems / Delays
Completion Date:	
Days Ahead of Schedute:	
Days Behind Schedule:	

Safely Issues:	Accidents/Incidents:

Summary Of Work Performed Today

Signature:	Name:

Equipment On Site	No Units	Working	
		Yes	No

Employee /Contractor	Trade	Contracted Hours	Overtime

Materials Delivered	From & Rate	Equipment Rented	No Units

Notes:

Date:		**Day:**	Mon	Tue	Wed	Thu	Fri	Sat	Sun
Foreman:									
Contract No:									

Hours Lost Due To Bad Weather	Visitors

Weather Conditions		Visitors
AM	PM	

Schedule / Problems / Delays

Schedule	Problems / Delays
Completion Date:	
Days Ahead of Schedute:	
Days Behind Schedule:	

Safely Issues:	Accidents/Incidents:

Summary Of Work Performed Today

Signature:	Name:

Equipment On Site	No Units	Working	
		Yes	No

Employee /Contractor	Trade	Contracted Hours	Overtime

Materials Delivered	From & Rate	Equipment Rented	No Units

Notes:

Date:		**Day:**	Mon	Tue	Wed	Thu	Fri	Sat	Sun
Foreman:									
Contract No:									

Hours Lost Due To Bad Weather	Visitors

Weather Conditions	
AM	PM

Schedule	Problems / Delays
Completion Date:	
Days Ahead of Schedute:	
Days Behind Schedute:	

Safely Issues:	Accidents/Incidents:

Summary Of Work Performed Today

Signature:	Name:

Equipment On Site	No Units	Working	
		Yes	No

Employee /Contractor	Trade	Contracted Hours	Overtime

Materials Delivered	From & Rate	Equipment Rented	No Units

Notes:

Date:		**Day:**	Mon	Tue	Wed	Thu	Fri	Sat	Sun
Foreman:									
Contract No:									

Hours Lost Due To Bad Weather	Visitors

Weather Conditions	
AM	PM

Schedule	Problems / Delays
Completion Date:	
Days Ahead of Schedute:	
Days Behind Schedute:	

Safely Issues:	Accidents/Incidents:

Summary Of Work Performed Today

Signature:	Name:

Equipment On Site	No Units	Working	
		Yes	No

Employee /Contractor	Trade	Contracted Hours	Overtime

Materials Delivered	From & Rate	Equipment Rented	No Units

Notes:

Date:		Day:	Mon	Tue	Wed	Thu	Fri	Sat	Sun
Foreman:									
Contract No:									

Hours Lost Due To Bad Weather	Visitors

Weather Conditions		
AM	PM	

Schedule	Problems / Delays
Completion Date:	
Days Ahead of Schedute:	
Days Behind Schedule:	

Safely Issues:	Accidents/Incidents:

Summary Of Work Performed Today

Signature:	Name:

Equipment On Site	No Units	Working	
		Yes	No

Employee /Contractor	Trade	Contracted Hours	Overtime

Materials Delivered	From & Rate	Equipment Rented	No Units

Notes:

Date:		Day:	Mon	Tue	Wed	Thu	Fri	Sat	Sun
Foreman:									
Contract No:									

Hours Lost Due To Bad Weather	Visitors

Weather Conditions

AM	PM

Schedule	Problems / Delays
Completion Date:	
Days Ahead of Schedule:	
Days Behind Schedute:	

Safely Issues:	Accidents/Incidents:

Summary Of Work Performed Today

Signature:	Name:

Equipment On Site	No Units	Working	
		Yes	No

Employee /Contractor	Trade	Contracted Hours	Overtime

Materials Delivered	From & Rate	Equipment Rented	No Units

Notes:

Date:		Day:	Mon	Tue	Wed	Thu	Fri	Sat	Sun

Foreman:

Contract No:

Hours Lost Due To Bad Weather	Visitors

Weather Conditions

AM	PM

Schedule	Problems / Delays
Completion Date:	
Days Ahead of Schedute:	
Days Behind Schedute:	

Safely Issues:	Accidents/Incidents:

Summary Of Work Performed Today

Signature:	Name:

Equipment On Site	No Units	Working	
		Yes	No

Employee /Contractor	Trade	Contracted Hours	Overtime

Materials Delivered	From & Rate	Equipment Rented	No Units

Notes:

Date:		**Day:**	Mon	Tue	Wed	Thu	Fri	Sat	Sun
Foreman:									
Contract No:									

Hours Lost Due To Bad Weather	Visitors

Weather Conditions		
AM	PM	

Schedule	Problems / Delays
Completion Date:	
Days Ahead of Schedute:	
Days Behind Schedute:	

Safely Issues:	Accidents/Incidents:

Summary Of Work Performed Today

Signature:	Name:

Equipment On Site	No Units	Working	
		Yes	No

Employee /Contractor	Trade	Contracted Hours	Overtime

Materials Delivered	From & Rate	Equipment Rented	No Units

Notes:

Date:		**Day:**	Mon	Tue	Wed	Thu	Fri	Sat	Sun
Foreman:									
Contract No:									

Hours Lost Due To Bad Weather	Visitors

Weather Conditions	
AM	PM

Schedule	Problems / Delays
Completion Date:	
Days Ahead of Schedule:	
Days Behind Schedute:	

Safely Issues:	Accidents/Incidents:

Summary Of Work Performed Today

Signature:	Name:

Equipment On Site	No Units	Working	
		Yes	No

Employee /Contractor	Trade	Contracted Hours	Overtime

Materials Delivered	From & Rate	Equipment Rented	No Units

Notes:

Date:		**Day:**	Mon	Tue	Wed	Thu	Fri	Sat	Sun
Foreman:									
Contract No:									

Hours Lost Due To Bad Weather	Visitors

Weather Conditions	
AM	PM

Schedule		Problems / Delays
Completion Date:		
Days Ahead of Schedule:		
Days Behind Schedute:		

Safely Issues:	Accidents/Incidents:

Summary Of Work Performed Today

Signature:	Name:

Equipment On Site	No Units	Working	
		Yes	No

Employee /Contractor	Trade	Contracted Hours	Overtime

Materials Delivered	From & Rate	Equipment Rented	No Units

Notes:

Date:		**Day:**	Mon	Tue	Wed	Thu	Fri	Sat	Sun
Foreman:									
Contract No:									

Hours Lost Due To Bad Weather	Visitors

Weather Conditions	
AM	PM

Schedule	Problems / Delays
Completion Date:	
Days Ahead of Schedute:	
Days Behind Schedule:	

Safely Issues:	Accidents/Incidents:

Summary Of Work Performed Today

Signature:	Name:

Equipment On Site	No Units	Working	
		Yes	No

Employee /Contractor	Trade	Contracted Hours	Overtime

Materials Delivered	From & Rate	Equipment Rented	No Units

Notes:

Date:		**Day:**	Mon	Tue	Wed	Thu	Fri	Sat	Sun
Foreman:									
Contract No:									

Hours Lost Due To Bad Weather	Visitors

Weather Conditions	
AM	PM

Schedule	Problems / Delays
Completion Date:	
Days Ahead of Schedule:	
Days Behind Schedute:	

Safely Issues:	Accidents/Incidents:

Summary Of Work Performed Today

Signature:	Name:

Equipment On Site	No Units	Working	
		Yes	No

Employee /Contractor	Trade	Contracted Hours	Overtime

Materials Delivered	From & Rate	Equipment Rented	No Units

Notes:

Date:		**Day:**	Mon	Tue	Wed	Thu	Fri	Sat	Sun
Foreman:									
Contract No:									

Hours Lost Due To Bad Weather	Visitors

Weather Conditions	
AM	PM

Schedule	Problems / Delays
Completion Date:	
Days Ahead of Schedute:	
Days Behind Schedute:	

Safely Issues:	Accidents/Incidents:

Summary Of Work Performed Today

Signature:	Name:

Equipment On Site	No Units	Working	
		Yes	No

Employee /Contractor	Trade	Contracted Hours	Overtime

Materials Delivered	From & Rate	Equipment Rented	No Units

Notes:

Date:		Day:	Mon	Tue	Wed	Thu	Fri	Sat	Sun
Foreman:									
Contract No:									

Hours Lost Due To Bad Weather	Visitors

Weather Conditions		
AM	PM	

Schedule	Problems / Delays
Completion Date:	
Days Ahead of Schedute:	
Days Behind Schedute:	

Safely Issues:	Accidents/Incidents:

Summary Of Work Performed Today

Signature:	Name:

Equipment On Site	No Units	Working	
		Yes	No

Employee /Contractor	Trade	Contracted Hours	Overtime

Materials Delivered	From & Rate	Equipment Rented	No Units

Notes:

Date:		**Day:**	Mon	Tue	Wed	Thu	Fri	Sat	Sun
Foreman:									
Contract No:									

Hours Lost Due To Bad Weather	Visitors

Weather Conditions

AM	PM

Schedule	Problems / Delays
Completion Date:	
Days Ahead of Schedule:	
Days Behind Schedute:	

Safely Issues:	Accidents/Incidents:

Summary Of Work Performed Today

Signature:	Name:

Equipment On Site	No Units	Working	
		Yes	No

Employee /Contractor	Trade	Contracted Hours	Overtime

Materials Delivered	From & Rate	Equipment Rented	No Units

Notes:

Date:		**Day:**	Mon	Tue	Wed	Thu	Fri	Sat	Sun
Foreman:									
Contract No:									

Hours Lost Due To Bad Weather	Visitors

Weather Conditions	
AM	PM

Schedule	Problems / Delays
Completion Date:	
Days Ahead of Schedute:	
Days Behind Schedule:	

Safely Issues:	Accidents/Incidents:

Summary Of Work Performed Today

Signature:	Name:

Equipment On Site	No Units	Working	
		Yes	No

Employee /Contractor	Trade	Contracted Hours	Overtime

Materials Delivered	From & Rate	Equipment Rented	No Units

Notes:

Date:		**Day:**	Mon	Tue	Wed	Thu	Fri	Sat	Sun
Foreman:									
Contract No:									

Hours Lost Due To Bad Weather

Visitors

Weather Conditions

AM	PM

Schedule

Completion Date:

Days Ahead of Schedute:

Days Behind Schedute:

Problems / Delays

Safely Issues:

Accidents/Incidents:

Summary Of Work Performed Today

Signature:

Name:

Equipment On Site	No Units	Working	
		Yes	No

Employee /Contractor	Trade	Contracted Hours	Overtime

Materials Delivered	From & Rate	Equipment Rented	No Units

Notes: